The Stars and the Black Holes

By Francisco Sanchez

The Stars and the Black Holes / Francisco Sanchez, Jr.

© **2021, Francisco Sanchez, Jr**

ISBN: 9798755129367

All Rights Reserved

No part of this book may be copied, sold or distributed, in either printed or electronic format, without the written permission from Francisco Sanchez, Jr.

Kindle Edition

The Stars and the Black Holes / Francisco Sanchez, Jr.

About The Stars and Black Holes

True Knowledge of How Existence Functions

Truly blessed is that man that has true knowledge of how existence or the universe truly functions, because that man also knows how that man himself functions.

And he also will do as existence herself to continue on existing or living on but existing or living on as if forever new and in complete, which is perfect, abundance, all five portions of her.

In The Stars and the Black Holes the fortunate reader will read or will know about the true origin of the stars, their purpose and also their end.

Also the fortunate reader will read or know about the Black Holes, their true purpose and their end.

The good book is very simple to understand because of not being technical or complicated and because of also having a very simple language for reading and understanding for the enjoyment of everyone.

Tags: stars, black holes, universe, cosmos, pioneer, frontier, auto-sufficient

The Stars and the Black Holes / Francisco Sanchez, Jr.

About this Author Beloved

All writers must come sooner or later to the things that they want to truly write about and most come to write where the easy money is and most writers make a very big killing in becoming very rich and very famous by writing fiction and good for them!

But the great question is how much fantasy for the human race since the human race has being living in a fantasy since the race entered into consciousness or began to think and invent with words?

Well, that is a great or tall question that Francisco Sanchez, Jr as a writer has truly asked!

And so he has truly chosen to walk or to actually write on the road less taken or less written about!

And so he began to research and write about immortality and as the same goes, one becomes what one thinks or writes or even reads about the most!

The Stars and the Black Holes / Francisco Sanchez, Jr.

Table of Contents

About The Stars and the Black Holes........pg. 3
About the Author........pg. 4
Prolog........pg. 6
Acknowledgement........pg. 7
Dedication........pg. 8

Introduction........pg. 9
Creation and the Number........pg. 14
The Stars and the Black Holes........pg. 23
The Small which makes the Great........pg. 29

Conclusion........pg. 32
Other Notes........pg. 40
Who am I?........pg. 48

Prolog

If you really want a miracle thus become a miracle and that miracle or oracle really begins with you when you confirm a higher mental conscious, which could be God as Creator or even your higher or superior self.

Now then, fortunate he that has true knowledge of how existence or how the universe really functions because now he knows how he himself really functions.

And he will also do as existence to continue existing or living but existing or living as if forever new and in complete abundance...

The Stars and the Black Holes / Francisco Sanchez, Jr.

Acknowledgement

I give thanks to my Father, who is beloved, for asking me once, something like 9 years ago, not to speak to him in a strange language because according to my father, he did not remember it to well!

Ever since I got interested in our rich English language and because of that I have become a glad writer!

Dedication

The book is dedicated to the good reader because the good reader is he who puts a period to what is written and, therefore, the good reader comes to complete the writer because now the writer is taller or greater.

Thanks, many thanks!

The Stars and the Black Holes / Francisco Sanchez, Jr.

Introduction

Before the beginning the universe was a neutral point, more as if a point zero even though the universe or the vacuum of empty space was in total darkness or in complete and deep darkness because of the simple lack of light or lack of matter or heat or even lack of true knowledge.

Also in the universe or in the vacuum of empty space there was not any movement or any attraction or friction or resistance not even time or hour or seconds or even a Nano second.

Neither there was a particle of dirt nor any gases nor even any weight at all, not even a negative weight or less.

The universe or the vacuum of empty space really was made of nothing, more as if of lack because there lacked the light, matter and even there lacked heat and time.

There was not even any weight positive or negative in the universe or in the vacuum of empty space, but nothing as nothing. What vanity!

Thus in truth, the universe or the vacuum of empty space was as if an empty or dead womb or something to avoid because of the enormous or deep lack which was the only thing which was.

But once light or matter or heat, which really was found outside of the universe, entered into the universe or into the vacuum of empty space, more as if into a womb, the universe or the vacuum of space was converted into one and also took the weight of one.

Now the universe or the vacuum of space was positive (+) because of being one or more…and time began to run or there began a time laps…

Light or matter had the form of a giant cube before entering into the universe or into the vacuum of empty space, which also was cold and dark, or without heat or without light.

Light or matter or heat also entered into the universe or into the vacuum of empty space with its 118 elements.

That is, now the universe or the vacuum of space was one composed of 118 parts and every part also had 118 parts.

Now then, light or matter or heat truly could be considered true knowledge or a true reality because of their numbers implanted or supplanted in them.

But the vacuum of space compressed the light or matter or element cube and turned it into a giant sphere or giant ball which exploded due to the enormous temperature caused by the compression of the vacuum of space or the universe.

Now light or matter was travelling throughout the universe or throughout the vacuum of space, which true form was sphere-like even though transparent or as if a foreskin really composed by a magnetic field.

But the weight of the universe now was less because of going down from one or more or less to 0.99999.

This new weight was because of the loss of energy as heat in the giant explosion…

Also the elements or the atoms were compressed and that way also causing an interior explosion of the elements or atoms and separating the electrons from the center or the nucleus and that way also converting the magnetic field into a negative field in where now the electrons rotated around the center or the nucleus of the atom.

After a long time, perhaps some millions of years, the smallest of matter or light stopped from illuminating for lack of light or energy and was converted into planets or moons or rocks in space.

Those planets which were attracted by the magnetic field of matter or stars still burning or illuminating thus those planets also became living wombs for different types of life forms because of the amount of chemicals or elements which those planets possessed.

After a long time, perhaps other millions of years, life began in those planets which converted their chemicals or elements into water and other components such as oxygen or other gases or liquids.

Thus life truly is a simple chemical reaction and the mind or consciousness or the brain also is a simple electromagnetic reaction, but with the suggested effort or the programming or even the command or word to duplicate existence, but which is not obligated to anyone or to anything but it is only a simple suggestion…

After some other millions of years, a lot of life under the waters began to seek refuge on dry land to survive and expand outside the waters.

Practically life under the waters took the physical form of new life or was physically transformed to be able to possess dry land or a habitat easy to control or to maintain with the gran purpose of life multiplying and also of surpassing not only physically but also surpassing mentally from the suggestion of multiplying or of duplicating existence.

After some other millions of years, there evolved the conscious beings but only one survived here on the earth.

The Stars and the Black Holes / Francisco Sanchez, Jr.

That survivor was man because of using his primitive conscious mind not only to survive but also to surpass or expand his mental consciousness...

But even though the universe has taken millions of years to expand and will also take many more millions to continue on expanding, the truth is that there is less matter or less light or stars and even though the universe will take millions of years to turn off or come to its end, to its end the universe will come...

Before the universe comes to its end, the universe will begin to convert into negative, (-). This will be caused by the giant stars when they convert into super nova or black holes when their nucleuses collapse in themselves because of their new and enormous weight.

These super nova or black holes will begin to suck in or to clean up as if vacuum cleaners all of matter and all of the stars with their planets and all life yet in existence in the universe or in the vacuum of space.

And once there no longer is any matter or stars or light in the universe or in the vacuum of space the super nova or the black holes will stop from existing because of lack of energy and they will disperse as if storms or hurricanes.

And the universe or the vacuum of space will return to what it once was, into nothing or into emptiness and there will not be any trace that there ever was a beginning before.

But there will be empty space for another or new beginning, but a beginning that will really become as if the only beginning...

To sum up, the universe had a beginning and the universe will also have its end. The force or the power which created the universe does not impose or will not impose to its end no matter how advanced life has become or the conscious beings...

The Stars and the Black Holes / Francisco Sanchez, Jr.

And the light which once illuminated, illuminated in vain. She served vanity until the end and for that reason she kept from illuminating or shinning and for that reason the light no longer is or will not be.

And as the same as the light the atoms with all their powers served vanity even though the electrons encircled or went around the center or the nucleus as if returning or as if trying to unite as before the beginning of the universe, such as ((-) (0) (+)).

But all of those efforts or movements only created friction and wear in the atoms and the diminution or reduction of matter or light from the stars or heat in space.

But if the conscious beings really do not want the universe to end or to come to its end, thus the conscious beings must do for that desire and take the position of the light and shine or illuminate for her or as if her so that the universe or the vacuum of space keeps refreshed for all times and never ever there be end or nothing or emptiness in space…

Creation and the Number

Creation really is about giving name or naming to start or to begin and giving rename or renaming to continue without beginning again even though to give rename or renaming is to become as if forever new and as if never there were a start or a beginning and neither end...

All of existence really is composed of numbers or be it positive numbers or be it neutral numbers or be it negative numbers or be it a combination of all the numbers at the very same time, but there will always be a number that is greater than the other negative numbers until that positive number is converted into a neutral number and after that positive number is converted into a neutral number thus also it will be converted with time into a negative number.

In other words, existence really is equal to (+ 0 -), but existence does not return to negative but rather a part of her and that part of her is the universe or matter or light, which returns to or really is converted from positive to neutral or to zero and much later it is converted to negative or into nothing.

The number zero is not negative or nothing. The number zero only is a neutral number which could be converted into a positive or negative number.

Existence as matter or as light is composed of three parts which add to 118 percent, 118 times 3 because they really are 118 positive parts, 118 neutral parts and 118 negative parts or (+ 0 -).

Now then, the neutral part or part zero is the universe and it is where it is added, it is subtracted, it is multiplied and it is divided at the very same time.

The positive part is the physical part of existence or the part from where comes out matter or light or the elements which truly are pure matter, but this part of existence appears to be minor than the other parts of existence.

The negative part of existence is the lack or is the vacuum of empty space or darkness which is composed of three parts, one is space or emptiness and the other two are space-time, which really are created by the interaction of the magnetic field between the positive and the negative.

When matter from the physical side of existence enters into the vacuum of empty space, matter even though a single piece, matter enters with its 118 parts assimilating the 118 elements.

That is, if only one element enters into the vacuum of space that element could assimilate or even can come to be converted into the other 117 elements...

But in reality matter enters from the physical side of existence into the vacuum of empty space in 118 pieces into 118 parts of the vacuum of space or into 118 dimensions at the very same time.

That is to say, into the vacuum of empty space there enter 118 elements and every element has 118 parts and at the very same time there enter 118 elements into 118 dimensions.

But that does not remain like that, because with every element the 118 parts are multiplied.

In other words, element number one has 118 parts but element number two has 236 parts and element number three has 354 parts and this pattern continues on until the last element, which is element number 118.

Also, the weight of the element is twice its number. That is to say, element number one has a weight of two and element number two has a weight of four and element number three has a weight of six. This pattern also continues on until element number 118.

Interestingly, all the parts of an element add to the number of that element. Thus, element number one has 118 parts and if we added those 118 parts thus we would get one.

That is, if we added 1+1+8 thus they would give us 10 and if we added 1+0 thus we would get 1.

In the case of element number two, if we added 2+3+6 thus we would get 11 and if we add 1+1 we would get 2.

This pattern continues on until element number 118. Element number 118 has 118 parts and every part also has 118 parts thus giving a sum of 13,924, also adding to one.

Now then, when element number one enters into the vacuum of space element number one enters with one proton, with one neutron and with one electron, (1+0-1).

And also element number one enters with a weight of two. But once in the vacuum of space, space compresses the element and the element is compressed causing the element to super heat and thus also causing the element to burst or explode.

In that burst or explosion the element loses a neutron or a third part and no longer the weight is of two but is less, such as 1.67. Also the electron was separated and now is spinning around the proton or the nucleus of the element, (+) -.

In the case in where an element has a high number thus that element will have neutrons in its nucleus, such as (+0+) -, -. Or (+0+0+) -, -, -.

This new transformation of the element, in where the electron is separated from the nucleus makes it possible for the element to be united to other elements and that way converting into a mixture of elements or isotopes…

Furthermore, before an element enters into the vacuum of empty space its three main parts have the same size or weight, ((+) (0) (-)).

But once that element enters into the vacuum of empty space thus that elements is divided into two parts, the nucleus in where new is the proton (+) and the neutron (0), and the outside part in where now is found the electron (-) going around the nucleus, ((+) (0)) (-).

And while the nucleus keeps its size or weight, the electron loses its size or weight because of the interaction or friction which it has with the nucleus or with the center of the element.

The interaction or friction which the electron has with the nucleus or with the center of the element also causes the electron to last less or lasts less time in the vacuum of space.

Once the electron is fused or is exhausted, the element or the nucleus is converted into a neutral element or without energy even though the nucleus is still positive or with protons and neutrons.

But the outside part or the electric field of the element is now a neutral field or the electrons have been converted into neutrons because of lack energy.

In other words, the seven electron rings of the element or the atom now are neutral when before they were negative.

And just as the element functions thus that way also functions the number and existence herself, but the number or the symbol of the number is only an illustration of the numbers but really does not show how is the number in existence or outside the vacuum of empty space in where there is no friction or movement even though there is a magnetic field.

Thus in truth, the number one or 1 outside the vacuum of empty space is represented by a cube.

Now, the cube or the number or the element one is composed of three main parts and they are the positive part, the neutral part and the negative part.

But those parts also are composed of 118 other parts. That is to say, that the positive part also is composed of 118 parts and the neutral part is also composed of 118 parts as the same as the negative part which also is composed of 118 parts.

And when the cube or the number enters into the vacuum of empty space the cube or the number is compressed into a sphere or into a globe but still with its 118 parts.

Thus, the number one is composed of not only 100 percent but also of another 18 parts or of another 18 percent. And if we added the parts thus we would have 1.

Now then, the number one or the symbol 1 as also is all of creation is a continuation because the start or the beginning is zero or a point or a neutral or an empty space.

But the number one or the symbol 1 also represents all of existence, the physical part, the neutral part and also the negative part.

Also the number one has the ability of converting itself into its 117 other parts also with their other 118 parts.

The Stars and the Black Holes / Francisco Sanchez, Jr.

In other words, the number or element one also has the ability of being infinite because also it could renew into a greater number such as the number two.

And that makes it possible the other 117 parts which add to 9 and the number 9 is a symbol of renovation. That first renovation extends the time of the number or of element one.

Thus, if we added the 117 parts which remain to one plus its other two parts, the neutral with its 118 parts, and the negative with its 118 parts, the sum would be of 353.

And if we added 353 thus it would give us 11 or eleven and if we added 11 we would get two, the possibility or the ability of the number or element one if it is renewed.

And once that the number or that element number one has renewed as two, thus the number or element one has become or will continue as double or for much more as two and as double the abundance.

And the very same step or process is with the number or with element number two. If we added all the parts that the number or the element two has, which are the double of one, thus it would give us four or 4.

That is to say, if we added all the parts of the number or element two thus we would get the double.

Now, the number zero or symbol 0 not only represents a reality but also represents a portion of existence as also represents peace or tranquility.

That is the reason that in many cultures the word no is use to attract peace, but the word is misinterpreted. Zero or peace also represents the beginning or start.

The number one represents continuation and also knowledge. And according as to how one was confirmed into the world thus one will do and one will hear and one would become as two.

That is to really say, the number two represents to hear or to do according to the knowledge or according to the confirmation given to one as one.

The number three represents unity or united to, thus according to what one has done with the knowledge given to one, one will be united as if three or a third part to him or her that gave knowledge or confirmed one because of one doing or because of one being born alive.

The number four represents adoration or praise because when one is united as if three or as if a third part thus one feels much gladness and much joy and one begins to sing or praise that third part.

The number five represents tests because of the praise which one has done. That praise which one has done because of feeling united with gladness and with joy thus will take one to tests to know if one truly was saying the truth or some confirmation or even a request or presentation for more or as more.

The number six is contention not only because of one being born and being born alive but also the number six is to contend for the consul of him or her that one praised but now has departed from one.

The number seven represents victory which could be granted to one with gladness and with joy and with the feeling of abundance as consul or reward because of the contention of one toward her, her meaning victory.

The number eight has much significance. The number eight not only represents all gladness and all joy but also the number eight represents the servant beloved of God or two hearts united as if one. Also, the number eight reflects or represents eternity or the very grandiose possibility of her or of entering into her as an immortal.

The number nine represents or means not only to overcome life or the world but also become as if new or renewed as savior beloved of God and with all the power and with all the authority of the holy heavens and also with all the power of riches and a step much closer to eternity or to immortality.

The number nine also represents auto-sufficient, which is a greater or higher mode of mental conscious or complete harmony.

The number ten represents or means dwelling, in where he that received the very grandiose or tall title of savior beloved of God could be the dwelling beloved of God and God also could be the dwelling beloved of the savior beloved of God.

And if the savior beloved of God accepts being the dwelling of God thus not only God also will be the dwelling beloved of the savior beloved but also the savior beloved will get even closer to eternity or to immortality.

In other words, the rolling or prodigal son or the electron has returned home or to dwelling or to the nucleus one more but this time for all times.

The number eleven represents the double abundance or the double plenitude, where the double abundance or the double plenitude of one is now the double abundance or the double plenitude of the other and the double abundance or the double plenitude of the other now is of one, thus the five portions.

In other words, eleven represents true abundance and because of eleven being true abundance thus eleven renews and eleven also is a step closer to eternity or closer to immortality or closer to the right side of creation or of God.

The Stars and the Black Holes / Francisco Sanchez, Jr.

And finally, the number twelve represents the right side or the eastern side of God or the right side of consciousness or the brain, which truly is an expansion of the conscious mind or a greater identity.

In reality, with every climb or raise of number thus one really has entered into a greater or taller mental consciousness.

And he that enters to the right side through confirmation of God or from his higher or taller consciousness thus not only is he the right side of God or from his higher or taller consciousness but also he has entered into eternity or into immortality...

To conclude, just as the number or the element in creation or in the vacuum of space truly is transformed or is converted into another number or into another element thus that way also the conscious being truly can be transformed or converted or enter into the next conscious or the next thinking mode or state.

And that transformation or that power of converting or that entrance into a greater identity truly is done though one's very own mouth...

The Stars and the Black Holes

The universe is a neutral point or is a point zero which is converted into a positive point or into a one or more when light or matter enters into the universe or into the vacuum of empty space.

The universe also is converted into a negative point when the grand majority of light turns off or the majority of matter no longer has energy…

Before light or matter enters into the universe or into the vacuum of empty space the weight of the universe is zero and when light or matter enters into the universe thus the universe takes on weight.

Light or matter is composed of 118 Elements and the weight of each element is two times its atomic number, more like its positive number.

In the case of element number one, for example, its weight is of two and in the case of element number two its weight is of four and in the case of element number 118 its weight is of 236…

Curiously, that the totality of the 118 elements adds to one and that the totality of their weight adds to 2.

That is to say, that if we added from one to 118 thus the sum would be 7,021. And if we added that sum thus it would be 10 and if we add that last sum thus it would be one. That is, 118 is equal to one…

And if we added from two to 236 the sum would be 14, 042. And if we added that sum thus it would be 11 and if we added that last sum thus it would be 2. That is, 236 is equal to two…

Thus, light or matter enters into the vacuum of empty space or into the universe as one or as a unit which is composed of 118 pieces or elements and the weight of the element is two times the atomic number of the element.

Thus in truth, number one, (1), itself is composed of 100 percent plus 18!

That is to say, that the number one or even oneself is equal or can become to 118 percent!

Also light or matter could enter into the vacuum of empty space or into the universe with only or as one element with its weight of two but that element would be able to convert into the other elements, even to the element 118 and its double weight of 236…

Element number one, for example, enters into the vacuum of empty space with its weight of two and it has 117 other possibilities of converting into the other 117 elements.

That is to say, element one is composed of 118 parts or pieces or the 118 percent and element number has 117 possibilities of converting into the other 117 elements according to the weight which element number one maintains.

In the same manner, element number two with its weight of four has 116 possibilities of converting into the other 116 elements or until the element number 118 with its weight of 236…

The Stars and the Black Holes / Francisco Sanchez, Jr.

When light or an element enters into the vacuum of empty space, light or the element enters as if it were a piece of magnet or as if it were a bar magnet,[(+)(0)(-)].

In the vacuum of empty space the magnet or light or the element is super compressed not only until it takes the form of a sphere or round but also light or the element or the magnet is super compressed until it gets to a very high level of temperature.

And when the temperature gets to its highest level thus light or the magnet or the element super explodes causing the light or the magnet or the element to divide into two parts and the part with less weight, such as the negative part, takes position in the magnetic field and that magnetic field now is a negative field, [(+)(0)] (-)…

The superior or the positive or the heaviest part of light or of the magnet or of the element takes position in the center or in the nucleus.

Thus, now we have the negative part of the magnet going around the positive part when before they were united and the center or the nucleus was neutral…

And even though the element lost weight because of the explosion or because of bursting, the element continues the one for two.

That is to say, its weight continues of two although the element one now is 0.999 and its weight is double, 1.998…

Thus, now element number one was reduced to about 0.999 with its new weight of 1.998 but to element number one also remains a neutron or even more than one or a neutral part which can be converted or can be transformed into a positive part and that way not only adding the number of the element but also adding its weight even though it will have only one negative particle going around the center or the nucleus or the positive side…

But if one or light or the magnet or the element does not convert into the next number or into the next element thus it loses its energy and will only be a piece of dead matter in the vacuum of space and it will be removed one day by the black holes…

Thus, light or the element is the same as a negative particle, is the same as a neutral particle and is the same as a positive particle.

In a way, one is equal to a negative portion plus a neutral portion plus a positive portion which totality is of 0.999 after entering the vacuum of space.

But in the vacuum of space light or the element is positive even though the vacuum of space is neutral but reacts as if negative because of the vacuum.

And when the element increases its positive part by converting the neutral part into a positive part, the element cannot attract the negative part because of the vacuum of space because now the negative part becomes as if more or its weight increases because of the weight that it receives indirectly from the vacuum of space.

And if there were no positive attracting the negative thus the negative would expand through the vacuum of space and it would stop being negative and it would be dead matter…

Now, an element, in this case a star, which number is high as the same as its weight thus lasts or remains longer in the vacuum of space or in the universe.

But the element or the star becomes heavier while the energy or matter to continue on lasts or it begins to transform from positive to neutral and once neutral, the element or the star practically becomes negative when its excess or super weight attracts the electrons toward the center and thus causing an implosion in where the element or the star becomes a nova or a new star but without light or without energy and that way causing an enormous hole in the vacuum of space when before the element or the star occupied the vacuum of space as an element or as a star…

In other words, matter or the star in the vacuum of space changes from positive to neutral and then from neutral to negative.

Thus, (+ 0 -), in where negative is really a black hole or a super vacuum cleaner in the vacuum of space.

This black hole practically eats or sucks all the matter around it to take all matter out from the vacuum of space or from the universe to make new or empty space for new matter or for another beginning…

But as long as there are black holes in the vacuum of space or in the universe, thus the vacuum of space or the universe is negative and as long as the vacuum of space or the universe continues as or is negative thus it keeps being for something and not for neutral and from neutral or from zero to positive…

Thus, so that the universe becomes neutral or to zero and from neutral or from zero to positive thus all the black holes or super space vacuums must stop from functioning and once the black holes or the super space vacuums stop from functioning for lack of matter or for lack of energy thus the vacuum of space or the universe will return to neutral or to zero…

Now then, this new vacuum in space or in the universe is neutral and has zero energy but even so attracts new matter or new stars or attracts the new or the next beginning which is outside of the universe or outside the vacuum of empty space…

And once new matter or new stars or new elements enter into the vacuum of empty space or into the universe thus the vacuum or the universe will be positive or one or more…

Thus, the cycle or the model or the rhythm of (+ 0 -) continues until the end of all times…

The Small which makes the Great

Existence is infinite or without size. Existence runs or she expands toward all sides at the very same time and at the very same time existence also compresses toward all sides, that way also adding to her very self an infinite weight and an infinite size.

That is to say, existence adds to herself, she multiplies herself, she divides herself and existence also subtracts her very self or from her very self to be able to become forever for more and even for new.

Existence forever existed and existence forever will exist. Existence has no time even though time existence is.

Existence is also composed of three principal parts which all add to one and that one is the grandiose part which makes the difference in all of existence.

Existence consists of matter, that which is physical. Existence also consists of empty space or of vacuum, that which is lack of something or of matter.

And existence also consists of movement or vibration or of that which many call space-time, which is a movement or vibration as if in waves in space, (((()))).

Thus in truth, existence is a simple magnet, [(+) (0) (-)]!

In other words, existence is composed of (+0-); in where (+) is equal to positive; in where (0) is equal to neutral; and in where (-) is equal to negative or the lack of.

In the scale of colors would be white, grey and black; in where white is light or matter; grey is neutral or space-time; and black is darkness or lack.

Now then, just as existence really is composed of three parts and practically of parts opposite to the other parts, (+0-), thus one also is composed of those very same opposite parts but in real knowledge because one really is real knowledge.

And one can really be positive knowledge or be neutral knowledge or even be negative knowledge.

But the greater the positive knowledge one has, less the neutral knowledge and less the negative knowledge or less the lack of negative knowledge.

In other words, the greater the knowledge of reality that one has, the greater the reality of one and less is the reality not known or less is emptiness or darkness.

Thus in truth, the more the knowledge of one, the more the abundance or the more is the life of one or more life makes sense to one.

Interestingly, that existence functions the same as one but existence knows it not but existence will never stop from being or will never stop from existing for all the eternity of eternity because she really makes herself for much more or renews!

That is to really say, even though all of existence is one, because of one existence really is and for all eternity infinite!

And every time that existence is increased by one or is increased by more than one, the point of neutrality or the neutral is less as the same as the negative or the unknown point is less.

But making the things less does not make existence greater, but making the things or existence greater makes existence greater!

And just as existence makes the things greater for existence to be greater herself, oneself can be greater by recognizing that life is more or that in life herself there is more and that there is with all peace, the neutral, and that here is with all knowledge, one or more than one; and also that there is with all gladness and with all joy and that also there is with all abundance of life, life renewed and in harmony or in auto-sufficiency…

Thus, through knowledge one renews for much more and because of one renewing for much more, one really can continue for all of eternity as the eternity herself…

Conclusion

Existence exists because she cannot stop from existing. Existence has no choice but to only exist.

Existence is composed of three portions of lack but which can be seen and existence also is composed of one portion which is but cannot be seen as of yet and which makes for three or even makes for four or for more portions…

Existence is composed of opposites which really are more opposites of themselves than to the opposites to which they are opposites of.

And this opposition or contrast which makes possible for existence to attract herself and it appears that existence moves or is in motion because of the very attraction of herself.

And existence is known or is seen more through the little which is seen of her and through the more which is seen of her, less of her is known or less is seen…

And even though existence exists and she attracts herself and she also is eternal knowledge and also she gives knowledge to the vacuum of empty space or to the universe and thus creating the beginning or the times, existence does not know that she exists even though she also renews herself so that she can continue existing and existing as if always new and as if nothing has ever happened and will never happen…

But existence renews herself or is reborn when the vacuum of space becomes empty again or when the knowledge given to the universe is not taken as acknowledgement and she once again gives or once again throws more knowledge into the vacuum of space or into the space of the universe, thus making as if a new beginning and everything which was before as if it never had being or as if it never were…

But that ability which existence has of renewing herself and of continuing for all eternity without knowing end or ending thus we conscious beings also possess.

But we are not obligated to live eternity as neither we are obligated to death or to the end or the ending.

Death or the end or the ending comes because it is not done to continue but to continue as if new and to continue forever or to continue for all the times in harmony or in complete abundance or in auto-sufficiency…

Those that do not want eternity thus they just wait for death and death will do for them so that there is no eternity for them…

But those of us that want to continue with life without seeing or without knowing death thus we must do to be reborn and that rebirth or being born again truly is done with knowledge or with acknowledgement.

Thus, we must give knowledge or give acknowledgement to that grandiose part of existence which is but which cannot be seen as of yet but even so it is what gives or it is what grants knowledge not only to the vacuum of empty space or to the universe but also it is what gives or what grants knowledge to every conscious being that requests it or that presents for it…

The Stars and the Black Holes / Francisco Sanchez, Jr.

And just as one presents oneself to that grandiose part of existence which gives or which grants or which even lends knowledge, thus that grandiose part of existence will be to one.

In other words, that grandiose or glorious part is the Creating part of existence or the renovating part or the part which gives identity to all existence or which reacts with all the opposites of existence...

Thus, he that gives or that grants or that lends to the Creating part or to the Renovating part of existence the knowledge or the acknowledgement of God or of Creator or of Renovator thus he also will have the knowledge or the acknowledgement of God or of Renovator and God will present to him or come to him or will allow him to draw near with that very same knowledge or acknowledgement...

Existence gives or grants or lends to the vacuum of empty space or to the universe knowledge in the form of matter or stars or light or even numbers, which take the conscious being to a greater or higher consciousness.

And if that knowledge in the vacuum of space or in the universe is converted into acknowledgement thus the universe will continue as if forever new without ever knowing end or ending nor knowing or remembering beginning...

But if that knowledge given or granted or lend to the vacuum of space or to the universe does not renew or is not converted or is not transformed into acknowledgement, thus that knowledge in the form of matter or stars or light thus will lose her energy and there will only remain and will be space dust no matter how large the piece of space dust and there will only be darkness or there will only be emptiness or vacuum...

The Stars and the Black Holes / Francisco Sanchez, Jr.

But that does not remain like that because now the dust which remained in the vacuum of space or in the vacuum of the universe must be taken out to give or to grant or to lend new knowledge to the vacuum of empty space or to the universe and that new knowledge will make a new beginning, in where there will not be any memory that there ever was a beginning before…

The manner in which existence takes the dust from the vacuum of space or from the universe is with black holes which truly are black vacuum cleaners.

The black vacuums or black holes also suck up any other matter or star which has remained in the vacuum of space or in the space of the universe…

Once there no longer is matter or dust in space thus the black holes also will turn of or will stop from functioning because of lack of energy and they will disperse or they will disintegrate in the vacuum of space.

This emptiness in space, which now has become as if a new vacuum or new space because of becoming as before the beginning thus will attract new knowledge in form of matter or stars or light…

And if in that new beginning the same happens which happened in the first, in where there was no renovation or there was no acknowledgement or no rebirth to continue, thus also will have its end or ending even though it may take billions of years…

But that does not have to be as the above because as long as there are conscious beings in the universe, the universe has the very grandiose opportunity of renovating or of rebirth or of receiving acknowledgement so that the universe because of the conscious beings the universe will continue without ever knowing end or ending…

The Stars and the Black Holes / Francisco Sanchez, Jr.

Existence without knowing it renews every time she lends knowledge in the form of matter or in the form of stars or of light to the vacuum of empty space or to the universe and the universe cannot continue for lack of acknowledgement or for lack of matter or for lack of renovated energy and thus the universe comes to its end or to its ending and thus making space for another beginning which will become as if the first beginning and also as if it never had an end or never an ending before…

But if the conscious beings are reborn or revive or take new life or receive acknowledgement through the very same knowledge or acknowledgement which they give or grant or lend, the universe will never ever have end or ending because the universe will continue as if forever new and as if it never had any beginning…

Now then, every conscious being truly has the very grandiose opportunity of being reborn or of reviving or of taking new life or of receiving acknowledgement or rename to be able to continue eternally with life and in complete harmony and in complete abundance…

But if the conscious being does not desire that very grandiose opportunity of living eternally and living in complete harmony and in complete abundance thus that conscious being only has to wait to die and that will he his end or his ending and nothing will become of that conscious being because eternity or immortality is not obligated or is not imposed, even a rock will stop from being or from existing…

Thus, one needs to be alive and conscious to be able to have or can receive immortality in the form of salvation and with her continue alive renewing and renewing also everything else as savior y protector or defender…

Just as knowledge in the form of matter filled with energy or in the form of stars or in the form of light enters the

The Stars and the Black Holes / Francisco Sanchez, Jr.

vacuum of empty space or enters into the universe, thus that same way also thought or knowledge or illumination enters the conscious mind.

That thought or that knowledge or that illumination can take the conscious being to a great state or from one state to another state or to a greater identity or from one identity to another identity even though that conscious being really continues with his physical form but every time that that conscious being enters into a greater state of knowledge or into a greater or new identity because of his knowledge, thus the physical form of that conscious being also is refreshed or is seen as if a new form...

But if the thought or if the knowledge or the new identity which enters in that conscious mind of that conscious being is a limited thought or is a limited knowledge or is a limited identity, thus that thought or knowledge or that identity, even though some type of energy or be it negative or be it positive, does not take that conscious being very far or into a greater state of identity, then that thought or knowledge or that identity will disperse and the conscious mind becomes once again as if empty.

And if that conscious being nothing does with his conscious mind to have thought or knowledge or identity so that the thought or the knowledge or the identity takes him to a greater state or to a greater identity in where not only his conscious mind will be refreshed or becomes as if a new mind but also his physical form or body also will refresh or even could be cured from certain lacks or faults, such as of that of deafness in one ear or both and also some emotional lack or fault such as loneliness or shyness and some other things...

But if the conscious being in the course of his life does not enter into a greater state of thought or of knowledge or of identity, thus the conscious being keeps on dying until he completely dies and his body will discompose until it turns to dust and the conscious being has lost his very grandiose opportunity of rebirth and of continuing with life as if new in complete or in perfect harmony and also in complete or in perfect abundance, perfect because it will be an abundance which will never ever end...

Now then, once the universe stops from discomposing or no longer the vacuum of space ever returns to nothing because of the conscious being coming to their maximum state or coming to their maximum identity and that way keeping the universe practically alive, thus there no longer will enter more knowledge in the form of matter or in the form of stars or of light because now that makes it possible the conscious beings because they will be the matter or the stars or the light or the illumination of the universe because of they being illuminated until the maximum or until perfection...

In other words, there will no longer be anymore beginnings nor there will be anymore ends or anymore endings and the universe will become as if there were never beginning because of the universe becoming as if new for all of eternity...

And all the different parts of existence will act or will react as if one as the same the body and the conscious mind of the conscious being will act or will react as if one or as a single part and existence will be one with the conscious being because of the conscious being becoming or as being existence herself and reflecting through his body her glory...

Thus, when one as seed for more came out from the entrails of a man and one entered into the entrails of a woman, one had no memory of that exit or entrance even though one as a seed was in harmony and in abundance in those entrails and one also came out with all gladness and with all joy and also with all feeling of abundance and entered into the entrails of a woman and there also sought for knowledge of life to life receive and in her also enter and once in her also one forgot because once again one entered in harmony while one was transformed or one took the form of life which one did for or for the one which one received through the act of one or because of one's movement or physical action to find life…

And when one became complete in those new entrails, one humbled and one took the grandiose position of contender and one came out or one entered into the entrails of the world not only as much more but also one came out or one entered for much more.

But in the world one did not remember that one came out from harmony while one keeps completing the form of contender which could take one to not only come to be conscious but also which could take one to rebirth and continue with life as if with a new form.

But if there were no rebirth because of lack of knowledge or because of lack of identity, thus that form not reborn would take one to death and that would be the end or the ending of all of her, life, and also of all of one…

Other Notes

Knowledge and Chaos

That which is not known or that which is not understood thus that is seen as if chaos even though with some form, but when that is known or when that is truly understood that takes form or takes reform and now is not chaos and now is as if it never lacked form…

Existence truly is all based upon knowledge or be it positive useful knowledge or be it negative or useless knowledge for being false.

There is not an existence completely positive or completely negative because existence exists because of existences being composed of opposite sides which show to be opposites to their very opposites but they are not because they themselves are not a single side or do not exist.

In other words, existence is composed of opposites which show that they are more opposites of themselves than the opposites they are opposites of but that composition is what really makes existence exist…

That is, darkness does not shine but darkness is seen. Darkness is not solid but darkness is seen and even though darkness is seen, darkness is lack, lack of light.

The same can be said about space. Space can be seen because of the lack, because of the lack of matter. And the same can be said about cold. Cold is felt for lack of heat or lack of warmth.

Thus darkness, space or cold is negative knowledge or is lack of knowledge. But lack does not make herself because lack is made by something which cannot be seen or is not known but that it really is…

In other words, the side that appears to be the right side can be the left side while the south side can be the north side because existence truly is a point of view and the greater the point of view, greater is existence or greater is her grandiose purpose…

Thus, existence truly is composed of four opposite sides which are set up one against the other creating attraction and that attraction also is composed of other opposites, such as positive and negative, which give movement to existence even though existence does not move because of her infinite size, but existence moves through knowledge so that more knowledge she can have as one once moved because of knowledge to have more knowledge so that one could be born and once in the world once again one moved or one did for more knowledge to be able to go on or so that one could continue with life…

Existence is one which truly is composed of infinite numbers or knowledge or be it positive or be it negative which the total of the sum forever truly will add to one, but that total will never be seen by eye but it will be known through understanding and once existence is truly understood by one thus through one existence will be seen because through one existence will reflect…

Thus, existence truly will be seen more because of her lack than her physical presence even though existence is infinite in size and in weight or in useful knowledge.

But even so without being able to be seen, existence can be truly known and understood because it is true knowledge and it is until the end of all time.

That is, existence will never have end because existence is true knowledge and because of being true knowledge, existence makes herself or moves because of knowledge even though existence already is complete and it is eternal…

Knowledge makes possible more knowledge and with that new knowledge existence truly is transformed as if taking a new form.

That in existence, such as matter or the stars or even the atom in the universe, which is not renewed or is not reformed because of lack of knowledge thus that will stop from functioning as such and will become dust and after becoming dust will also stop functioning as dust because it will be separated from all its sides or opposite elements which make possible its existence…

Thus, obviously, the conscious or the living beings can imitate more of existence than existence herself and the very particles that come out from her.

And even though the stars last a lot of time in the vacuum of space or in the universe, the stars with time will come to lose their function as stars and begin to turn off.

The stars stop from functioning because they do not have from where to grab from or from where to take out from them very selves to continue functioning as stars because the stars no longer have the matter or the knowledge in themselves to reform or to transform or to renew themselves…

But the conscious or the living being can truly take out or can truly grab knowledge, which really is matter, from the conscious or the living being him very self just as existence

The Stars and the Black Holes / Francisco Sanchez, Jr.

her very self or the physical part or the positive part of existence does.

And that truly has to do everything with peace and also with knowledge. Before the beginning of things there was in the vacuum of empty space a very profound peace, so profound that it attracted light or matter or knowledge or illumination and thus breaking the very profound peace because the vacuum of space began to compress the light, the matter, the knowledge or what was illumination and that way also creating as if chaos because not only the vacuum of space was no longer the same vacuum or emptiness but also the form of the light or of the matter or of the knowledge or the form of that which was illumination was no longer the same form because the light or the matter or the knowledge or the illumination took another form or reform…

Well then, imagine, if it is that one can imagine, when thought or illumination entered for the very first time into the mind of man and man did not know how to react but he felt as in chaos or in desolation because of lack of peace and because of lack of understanding of that new state of man.

But even so man never again remained as a simple animal because now man had a new identity, the identity of a conscious or living being.

And with that new identity man could surpass all the times if man wanted to surpass all the times or to live beyond the times because of the new form or reform of man…

But for man as conscious or as living being can surpass the times, and surpass means taking all the other forms or reforms or states of consciousness or of thought which still remain to be able to continue with life, thus man must do again for his knowledge as once man did for knowledge before being born and to be able to be born and thus take the form or the reform or that state of knowledge or of that new identity...

But if man after coming to a new form of knowledge or of identity does not do to come or to get to the next form or reform or new identity, thus man begins to lose his actual form and man dies and man stops from existing as also all the other things in the vacuum of space or in the universe stop from existing because of lack of knowledge or because of lack of new form or reform or even for lack of new identity because everything which stops from existing, even a rock, stops from existing because of lack of new form or because of lack of new identity...

Thus, just as physical existence renews through knowledge because of existence truly being knowledge and that way complete existence taking as if a new form or reform or new identity thus man also for being knowledge through knowledge man also can take as if a new form or reform or a new identity...

But if that new knowledge of man truly begins or is in the conscious mind of man, which gives man identity as man or as conscious or living being and she was granted to man because of some good action or good act of man such as a good feeling of peace or of feeling that in the world as in existence there was more of what man himself could see or could imagine...

In other words, man was illuminated because of his search or because of his good feeling that there was much more in the world as also in existence her very self.

That illumination, which truly is a very grandiose expansion of conscious mind, was granted by the brain or by Him that many of us call God as a form of peace or of consul or as a form of given knowledge or of giving man a new identity of conscious or living being.

And once man achieved that new identity of conscious or of living being, man could share that new knowledge with the rest of mankind so that the rest of mankind also entered into mental consciousness or that the rest of mankind could be granted the conscious mind.

But that did not remain like that, because now all the sons born after the illumination of man or after man entering into the conscious mind thus also the sons were born conscious or living beings if the sons were born alive...

Thus, when one enters into a new illumination or into a new knowledge or into a greater state of conscious mind or into a new identity, thus one once again must do a new action of confirmation or presentation or even of request, but for much more.

But, obviously, this new or this next confirmation or proclamation or even request truly has to be greater or truly superior to the confirmation or the proclamation or even greater or superior to the request of before because one truly will enter for better or for greater into a better or into a taller or greater illumination or into a better or greater knowledge or into a better or greater state of conscious mind or into a new identity which surpasses the former identity...

One truly will know that one has entered into illumination or into a greater state of consciousness when one feels harmony of profound peace, of gladness, of joy, of abundance and one also hears a voice offering or proclaiming such thing as that one will lack nothing or something similar but better or greater than the first illumination or better than the first state of consciousness, in where one truly did not know what was happening to one but even so one became for much more for one truly entering into a greater state of mental consciousness...

But because of one entering into that new state of mental consciousness thus one entered as if into chaos because of one not having the peace and not having the knowledge of that new state to be able to take the new form or the new identity of one and come out of what seems chaos.

And just as one truly entered or went up into a new mental consciousness state thus also the brain or He who one called God entered or went up into a new state of mental consciousness or entered or went up into a new heaven in where also peace is lacking and knowledge is lacking to be able to take His greater form or reform or His new identity as Creator and come out with all power and with all authority from what seems to be chaos…

When one was born to the world those that received one comforted one and one rested and one began to adapt to the world while one grew and also one was comforted by those that received or adopted one as son until one dominated or one became accustomed to or one adopted the world and also took the identity with which one was born with.

But when one enters into a greater state of mental conscious the world can no longer comfort one because the world does not have that peace and does not have that knowledge to complete that new state of knowledge or new identity and the peace that also comes with that knowledge or new state or new identity…

Thus, that is the reason that one has to look up as if into the sky or one has to seek as if in the heavens so that one can truly find or one can truly receive knowledge so that one can enter or go up into the next thinking mode or state of mental consciousness or into a greater or taller identity or form or reform…

When one was born, one truly was born with the form or reform or with the identity of contender or with that identity of contention and so one triumphed without one knowing and one grew until entering into consciousness or entering into understanding or entering into a new mind.

Thus with that new mind one has to do to continue entering into more consciousness so not to lose the form or the reform or the identity of one which truly continues to be of contender, of contender because one truly continues struggling or contending for peace and for knowledge because of one being a conscious being and the conscious being are formed or are reformed or enter into a greater consciousness through knowledge or through acknowledgement but through knowledge or through acknowledgement granted by his very request because what truly grants one's request is the brain or that or He which one calls God or the God of one or even the God of the fathers of one…

Now, with every grant of peace and of knowledge one truly enters into a better or the brain or that which we call our God expands our mind and that expansion or that new state or that new identity truly fills one with gladness and with joy and one also feels much abundance.

But that is not all, because one also hears a voice confirming one's request.

Many times the voice is even heard before one hears the confirmation and one becomes filled with gladness and joy and one also feels with a good sense of abundance or of presence of life even though the abundance or the presence of life is not seen.

But all of that feeling of gladness and of joy and of abundance truly is a very good indicative that there is still more and that the next request will be rewarded also through gladness and through joy and also in all abundance of life or new life…

Who am I really?

My pen names as a writer are Francisco Sanchez and Forester de Santos and I am on a very grandiose or tall crusade of rebirth alive or to be born again with complete gladness and with complete joy and also with complete abundance of God but as much more than God and as much more than Creator.

Now then, one who truly is on a very grandiose crusade cannot follow another or cannot let himself be surrounded by his beloved ones or his fans because he cannot cross over them or he cannot cross over because of them being in the way or because of them blocking the path which is but which cannot be seen until rebirth or until one is born again.

I do not ask to be followed, not because I will not lead, but because I will not look back but I will look to my right and to my left to see who walks with me.

But those that truly decide to follow me will become as me and as me will truly receive or gather true knowledge because my struggle or contention or my very grandiose crusade of rebirth is true, so true in fact that I have become a much better person because of the true faith which I have come to receive through my search and research for the truth.

And because I have come to have true faith or faith of God, thus I use my true faith as a shield to repel or to reject other beliefs or good sounding lies!

Therefore, to rebirth alive or to be born again while still living here on the very earth which will be as in the very heavens through rebirth or through auto-sufficiency!

Now then, God could be the Master Creator, the brain, the sub-conscious mind, the conscious mind surpassed into a taller consciousness or even much more, but all of these forms or reforms of God or of life forever will be every time higher in consciousness and so that man could enter or overcome a higher conscious thus man has to present himself for more or as more to what is higher than man so that which is higher than man, God, allows man to go up to the higher or to the next mode of thinking or of mental consciousness...

The Stars and the Black Holes / Francisco Sanchez, Jr.

$$(- (- (- (- (- (- (0 + 1) +) +) +) +) +) +)$$

If you truly enjoyed this simple and humble work, please leave a comment according to your good pleasure and give also a rating but also according to your good pleasure.

Thanks so very much for your time and best of wishes, Francisco Sanchez, Jr.

Thanks for reading my work!
0+1 = peace and knowledge to all mankind…

The Stars and the Black Holes / Francisco Sanchez, Jr.

These pages are for your personal use

The Stars and the Black Holes / Francisco Sanchez, Jr.

These pages are for your personal use

www.ingramcontent.com/pod-product-compliance
Lightning Source LLC
Chambersburg PA
CBHW050314220526
45465CB00005B/1980